THIS BOOK BELONGS TO:

_____

THE WONDERFUL WORLD OF RABBITS

Mimi Jones

Dedicated to all the rabbit lovers.

ISBN 978-1-958985-44-1

Text copyright © 2025 by Mimi Jones

www.joeysavestheday.com

A Mimi Book

WELCOME TO THE WONDERFUL WORLD OF RABBITS

3
Foot

Rabbits are great jumpers! They can leap up to 3 feet high and 10 feet long.

6

POWERFUL

They have powerful hind legs that help them escape from predators quickly.

Baby rabbits are called kits or kittens.

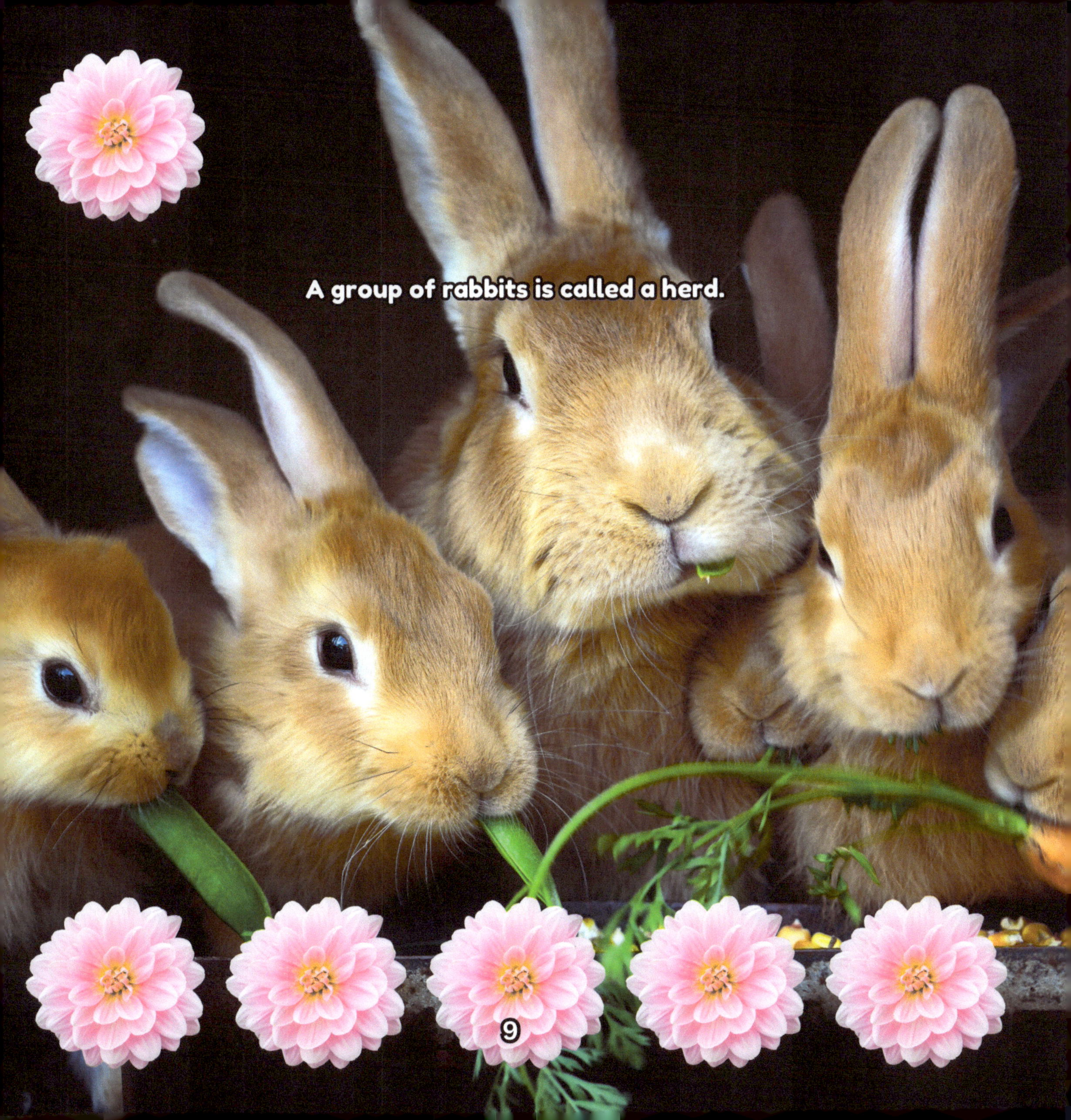

A group of rabbits is called a herd.

Rabbits are herbivores, which means they only eat plants.

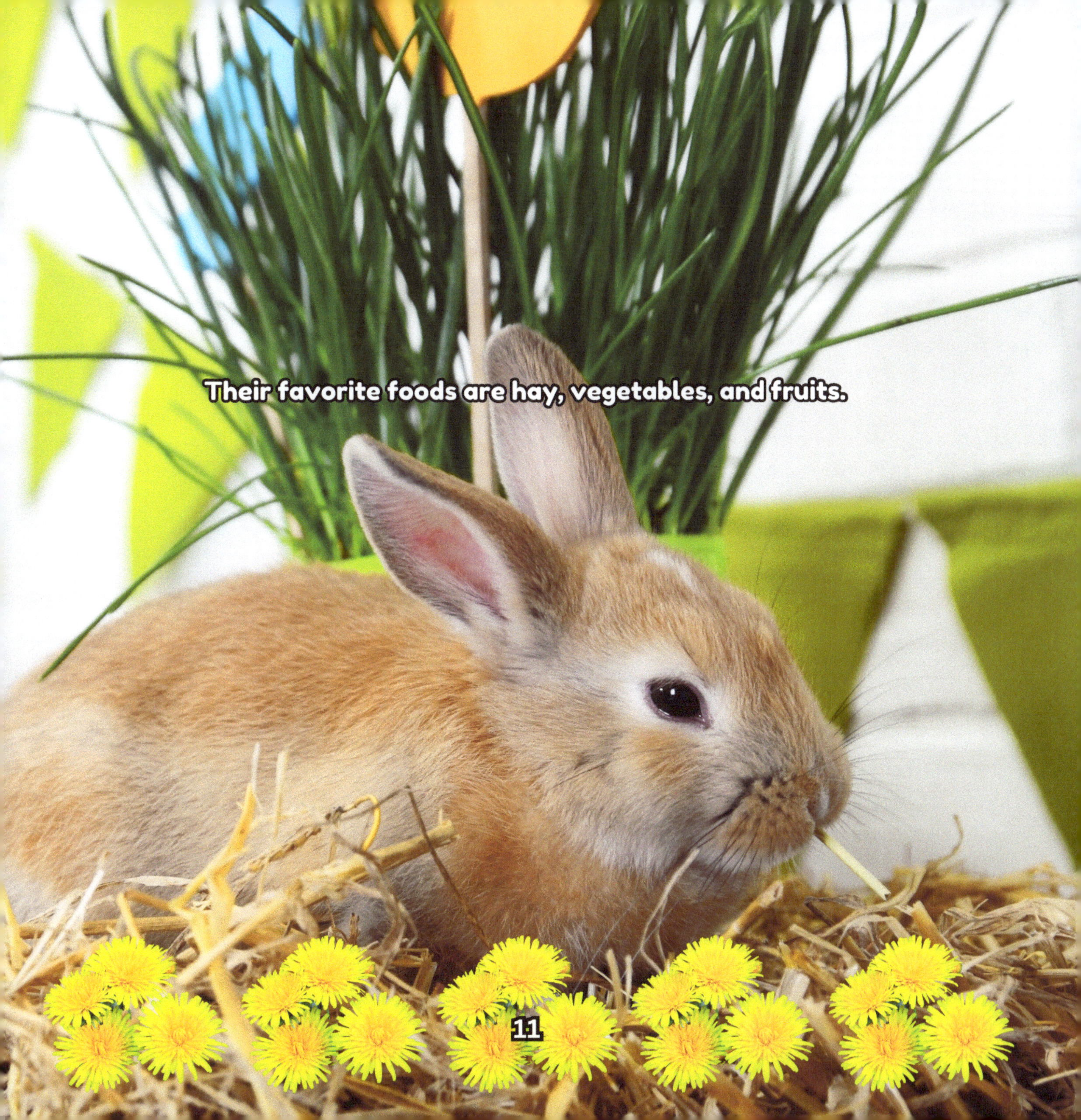

Their favorite foods are hay, vegetables, and fruits.

28

Rabbits have 28 teeth, and their teeth never stop growing.

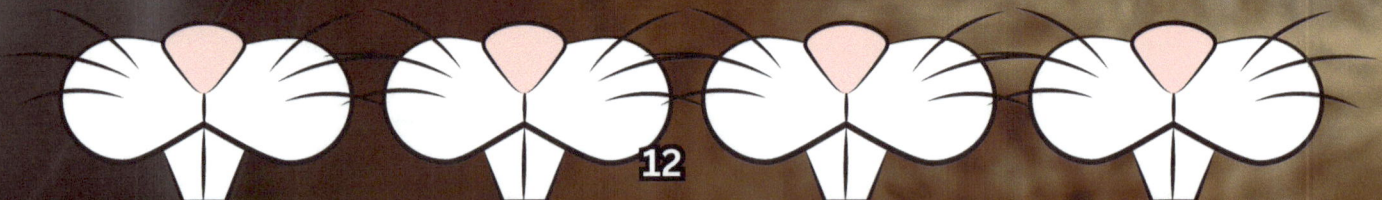

They use their whiskers
to feel their way around
in the dark.

180°

Rabbits can turn their ears 180 degrees to listen for danger.

They have the ability to see what's behind them without needing to turn their heads.

SOCIAL

They are very social animals and like to live in groups.

Rabbits live in burrows called warrens.

Rabbits communicate by thumping their hind legs.

They have a special way of grooming
themselves with their paws and tongues.

Rabbits have a lifespan of about 8 to 12 years.

Active

They are primarily active during the early morning and evening hours, a behavior known as being crepuscular.

Rabbits have a unique way of digesting their food. They produce soft droppings called cecotropes and eat them to get extra nutrients.

Excellent!

A rabbit's hearing is excellent, and they can hear sounds from very far away.

SNIFF

They have a keen sense of smell that helps them find food and detect predators.

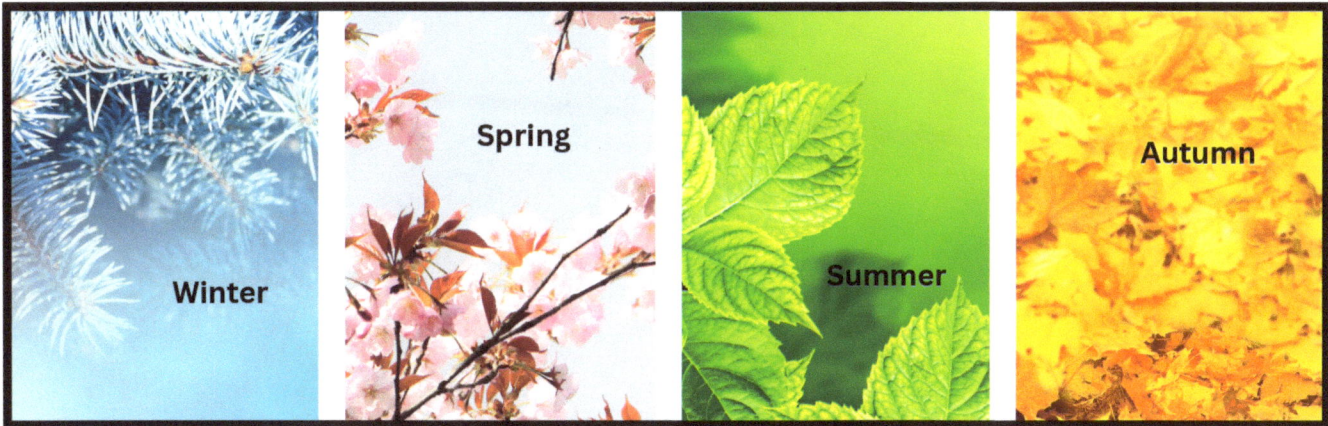

Winter

Spring

Summer

Autumn

Rabbits shed their fur regularly and grow new coats for different seasons.

the Great escape

They can run up to 35 miles per hour when they need to escape danger.

35 mph

Rabbits dig extensive tunnel systems to create safe homes.

They are very clean
animals and groom
themselves several
times a day.

See you!

Rabbits have large eyes that allow them to see well in low light.

A rabbit's tail is called a scut,
and it's very short and fluffy.

30

Rabbits are found all over the world, except for Antarctica.

North America

Europe

Asia

South America

Africa

Australia

Antarctica

31

Count the rabbits.

Thank you for taking
the time to read this.
I hope you found it
informative and gained
some valuable insights.

THE END!

Check out these other interesting books in the Wonderful World of series!

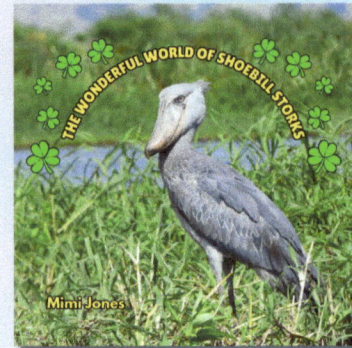

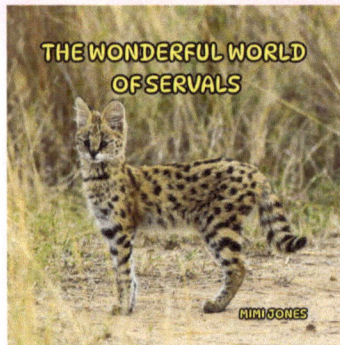

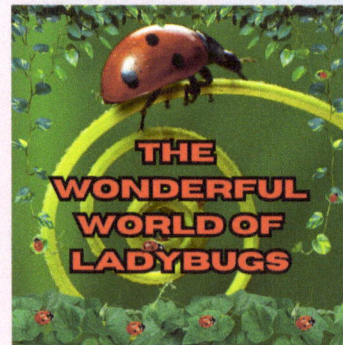

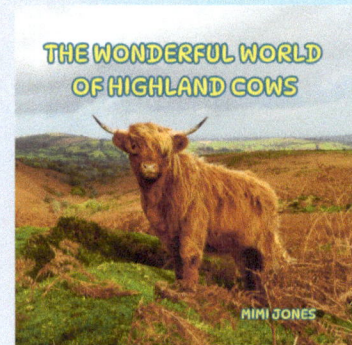

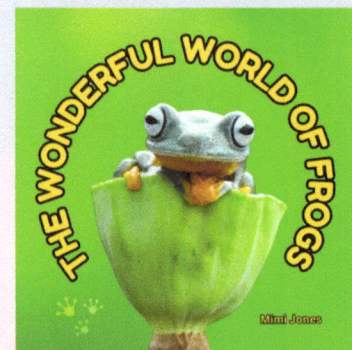

THE WONDERFUL WORLD OF SUNFLOWERS

THE WONDERFUL WORLD OF DRAGONFLIES
Mimi Jones

THE WONDERFUL WORLD OF SHOEBILL STORKS
Mimi Jones

THE WONDERFUL WORLD OF SERVALS
MIMI JONES

THE WONDERFUL WORLD OF LADYBUGS

THE WONDERFUL WORLD OF HIGHLAND COWS
MIMI JONES

THE WONDERFUL WORLD OF PANDAS
Mimi Jones

THE WONDERFUL WORLD OF BEARDED DRAGONS
Mimi Jones

THE WONDERFUL WORLD OF FROGS
Mimi Jones

www.mimibooks.com

www.ingramcontent.com/pod-product-compliance
Lightning Source LLC
Chambersburg PA
CBHW060835270326
41933CB00002B/96